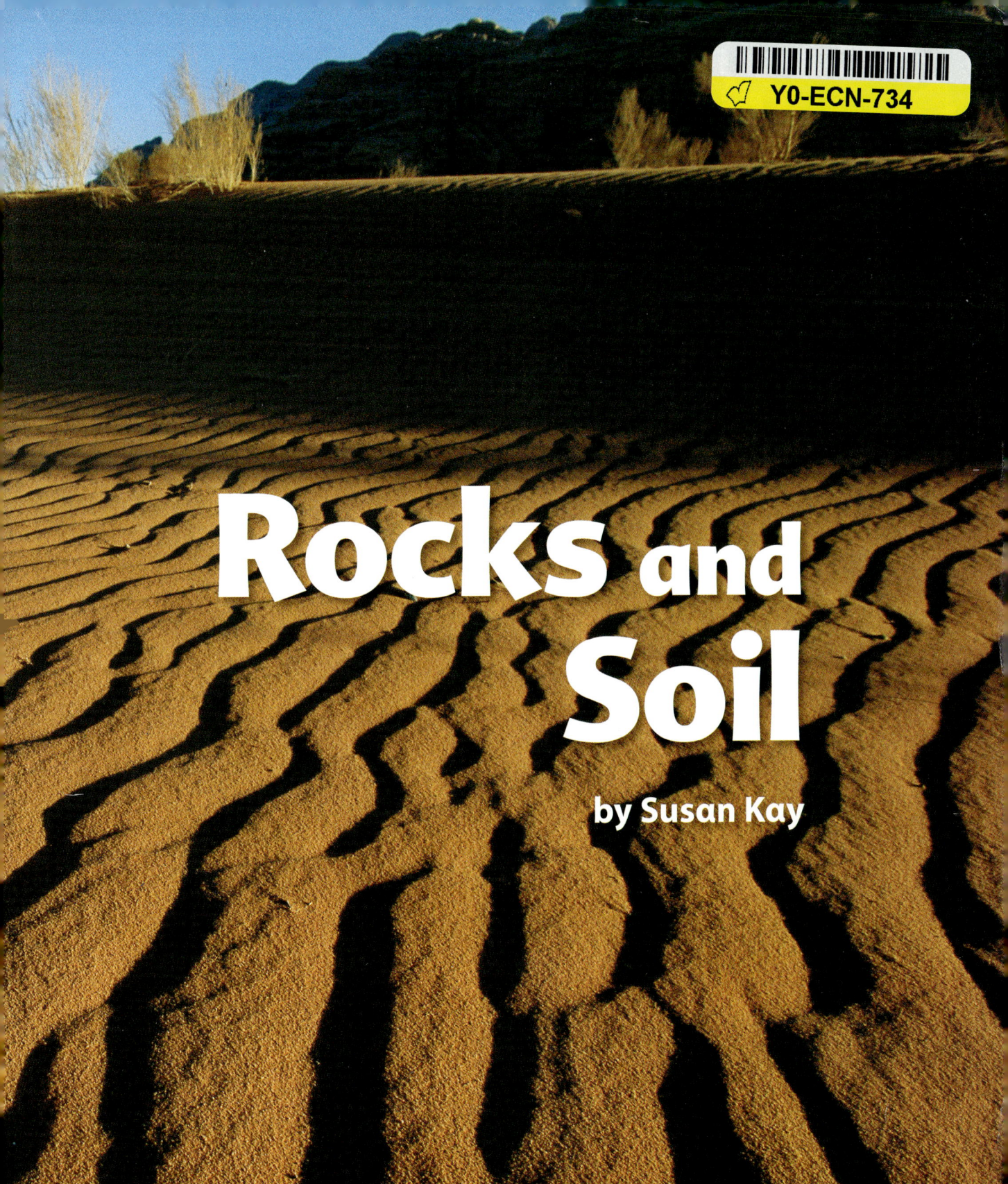

Rocks and Soil

by Susan Kay

2

Contents

Introduction: Rocks and Soil in Nature 4

Chapter 1

Big Idea Question

What Can You Observe About Rocks? 6
   Properties of Rocks 8
   People Use Rocks 10

Chapter 2

Big Idea Question

How Do Rocks Change Shape? 12
   Weathering 14
   Ice and Wind 16

Chapter 3

Big Idea Question

What Can You Observe About Soil? 18
   How Soil Is Formed 20
   Properties of Soil 22
   People Use Soil 24

Conclusion: Earth's Rocks and Soil 28

Glossary 30
Index 32

Introduction

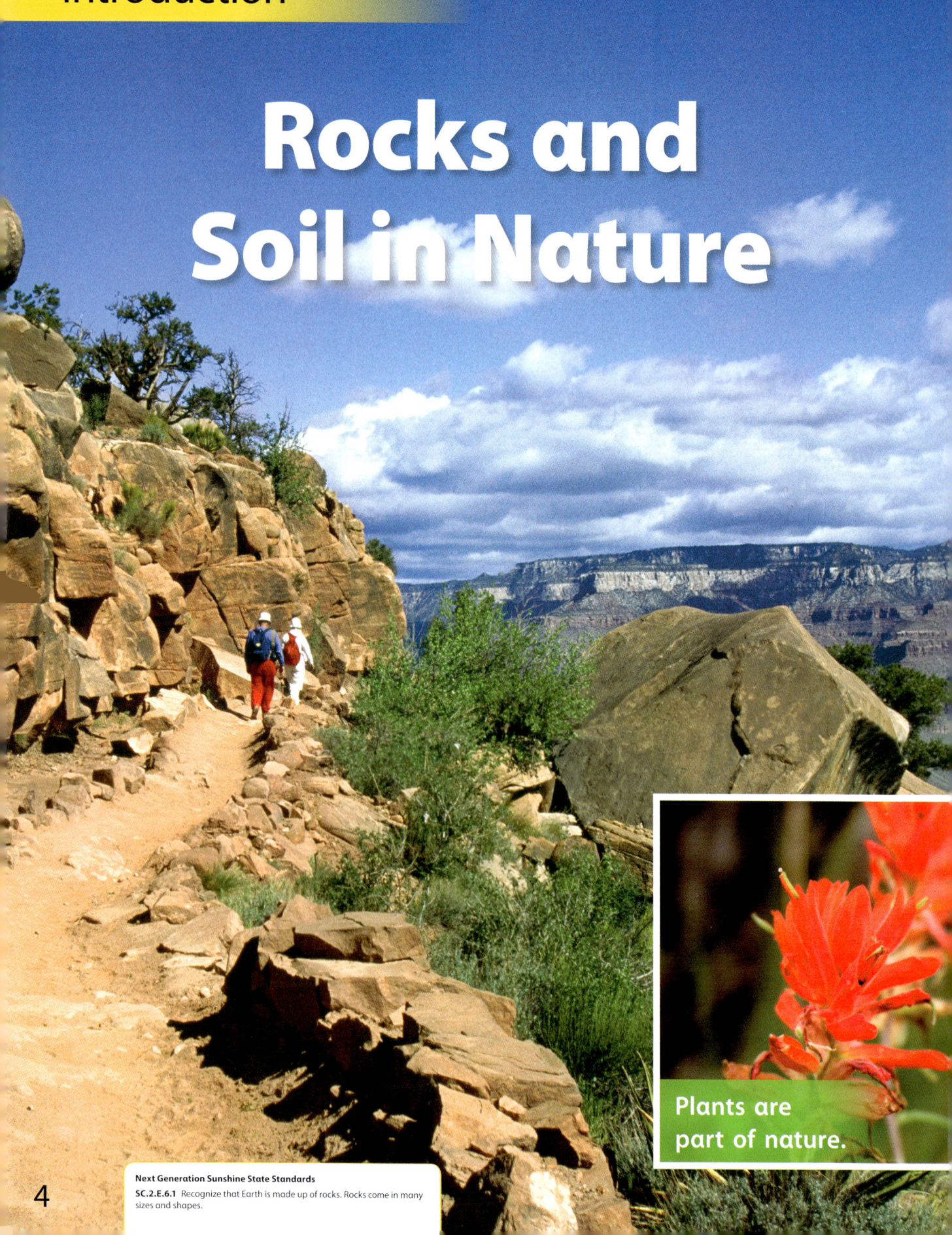

Rocks and Soil in Nature

Plants are part of nature.

Next Generation Sunshine State Standards
SC.2.E.6.1 Recognize that Earth is made up of rocks. Rocks come in many sizes and shapes.

Imagine hiking on a nature trail. You see plants and animals. You also see rocks, soil, and water. Rocks, soil, and water are part of nature.

Rocks and water are part of nature.

Chapter 1

Big Idea Question

What Can You Observe About Rocks?

Next Generation Sunshine State Standards
SC.2.E.6.1 Recognize that Earth is made up of rocks. Rocks come in many sizes and shapes.

Rocks are almost everywhere on Earth. There are many kinds of rocks. Rocks come in different shapes and sizes. Rocks have different colors, patterns, and textures, too.

Properties of Rocks

A **property** is something about an object that you can observe with your senses. For example, shape, size, color, and **texture** are properties. Texture is how an object feels. You can sort rocks by their properties.

What can you observe about the properties of these rocks?

Properties	Describing Words
• color	• white, tan, brown, and so on
• texture	• rough, smooth
• layers	• layered, not layered
• fossils	• fossils present, fossils absent
• air holes	• air holes present, air holes absent
• strength	• crumbly, strong

People Use Rocks

People use rocks to build many things. This famous monument in Washington, D.C., is made mostly of white marble.

Rocks were used to build the tall, white Washington Monument.

Some rocks are cut into different shapes and sizes and used to make buildings. Other rocks are cut and polished to make jewelry.

People use machines to get rocks from underground.

This building is made of rocks.

Polished rocks are put into this piece of jewelry.

11

Chapter 2

Big Idea Question

How Do Rocks Change Shape?

Next Generation Sunshine State Standards
SC.2.E.6.1 Recognize that Earth is made up of rocks. Rocks come in many sizes and shapes.

Many rocks are very hard. It seems like they could never change. But rocks do change. They change shape. This happens slowly, over time.

Water, wind, and ice change the shape of rocks. They slowly break down rock, or wear it away until it becomes part of the soil. This is one kind of **weathering**.

Weathering

Water in a river flows over rocks day after day and year after year. The water slowly wears away the rocks. It changes rough rocks and makes them smooth. This is one kind of weathering.

Water weathers these rocks in a river.

There is also another kind of weathering. It happens when water dissolves all or parts of rocks.

Parts of these rocks have dissolved and changed shape from weathering.

Ice and Wind

Ice weathers rocks, too. Water gets into cracks in rocks and freezes into ice. The ice pushes on the rock. Over time, rocks break apart and change shape.

Ice caused this rock to change shape by cracking.

Wind also weathers rock. Sometimes the wind picks up sand. It blows the sand over rocks. The wind wears rocks down by rubbing them with sand.

Wind picked up small pieces of rocks and sand. It wore away rock to make this shape.

Chapter 3

Big Idea Question

What Can You Observe About Soil?

Next Generation Sunshine State Standards

SC.2.E.6.2 Describe how small pieces of rock and dead plant and animal parts can be the basis of soil and explain the process by which soil is formed.

SC.2.E.6.3 Classify soil types based on color, texture (size of particles), the ability to retain water, and the ability to support the growth of plants.

Bits of rock can become part of **soil**. Soil is a layer of loose material that covers much of the land on Earth's surface.

How Soil Is Formed

Bits of rock, air, and water are parts of soil. They are nonliving. The small bits of rock in soil come from the weathering of bigger rocks. **Humus** is part of soil, too.

Humus is usually dark and crumbly.

Humus is bits of living things, such as plants and animals, that have died and decayed. Humus mixes with bits of rock, water, and air. Soil forms when these parts mix.

The top layer of this soil has a lot of humus.

Air and water fill spaces in the soil.

Worms and decaying, or rotting, leaves can enrich the soil.

Properties of Soil

There are different kinds of soil. Each kind has different properties. Color and texture are two properties of soil.

Soil that has a lot of clay is sticky when it is wet.

Sandy Soil	Clay Soil	Humus Soil
This soil has a lot of sand in it. It feels like a handful of sugar. It is usually light brown.	This soil has a lot of clay in it. It feels soft and sticky. It is often reddish brown.	This soil has a lot of humus in it. It feels soft and crumbly. It is usually dark brown.

People Use Soil

People use soil for many things. People use soil to grow plants for food. Some kinds of soil are better for growing food than others.

Clay soil holds too much water for most plants. It is not good for growing most food.

Some plants can grow in sandy soil. But sandy soil is not good for growing most food. It does not hold enough water.

This soil has a lot of humus. It is good for growing plants. It holds the right amount of water for most plants.

25

People use soil to make pottery, tiles, and china. Building materials are made from soil, too. Bricks are made from clay. Concrete is made with sand.

These colorful pottery jars are made of clay.

People also use soil for building. Soil is the firm base under many homes and roads.

A new house will be built on this soil.

Conclusion

Earth's Rocks and Soil

Rocks have properties, such as size, shape, color, texture, and layers. People use rocks in many ways.

Weathering changes rocks. Wind, water, and ice can wear down rocks or break them apart. Water can dissolve all or parts of rocks.

Soil is made up of bits of rock, air, water, and humus. People use soil in many ways. People grow food in soil. They build homes and roads on soil. They make building materials from soil.

Next Generation Sunshine State Standards

SC.2.E.6.1 Recognize that Earth is made up of rocks. Rocks come in many sizes and shapes.
SC.2.E.6.2 Describe how small pieces of rock and dead plant and animal parts can be the basis of soil and explain the process by which soil is formed.
SC.2.E.6.3 Classify soil types based on color, texture (size of particles), the ability to retain water, and the ability to support the growth of plants.

Glossary

humus (page 20)

Humus is a part of soil. It has bits of decayed plants and animals.

Many plants grow well in **humus.**

property (page 8)

A **property** is something about an object that you can observe with your senses.

Having layers is one **property** of rocks.

soil (page 19)

Soil is a layer of loose material that covers part of Earth's surface.

Most plants grow in **soil.**

texture (page 8)
Texture is the way an object feels.

Some rocks have a rough **texture.**

weathering (page 13)
Weathering is the breaking apart or dissolving of rocks.

Weathering can cause the shape of rocks to change.

Index

clay 22–24, 26

fossil 9

humus 20–21, 23, 25, 28, 30

property 8–9, 22, 28, 30

sand 23, 26

soil 19–28, 30

texture 7–9, 22, 28, 31

weathering 13–17, 20, 28, 31